A Shocking Journey

Stephen J. Wolf, PhD

ISBN: 0-9969846-4-X
ISBN-13: 978-0-9969846-4-5

To the 1,500 (and counting!) students I've had the honor and joy of teaching, this work is for you; for inspiring me everyday, for asking questions, for keeping me on my toes, and for always keeping me young.

Thank you for everything.

TABLE OF CONTENTS

CHAPTER 1

AN ORDINARY DAY

It was a typical Monday in Merl-Mont Middle School as Nathan entered the lunchroom and sat with his friends. The tables were still moist from being cleaned between periods and even though he had just arrived, the room was full of excited noise.

"I guess it's French fry day," Nathan grinned, pulling out a brown bag with a peanut butter and jelly sandwich.

His best friend, Leena, nodded and shrugged, pulling out her own lunch. "Don't get me wrong. I like French fries like everyone else, but the ones they have here are always so hard to chew."

Reiko smiled, her face lighting up as she did so. "Fries at the bowling alley are the best. They're the only ones I eat."

There was a ruckus at the fries station, after which Sergei walked away feeling dejected. "Rats."

"Something wrong?" Nathan asked.

Reiko looked at Sergei's lunch tray. "Where's your sandwich? Or an apple? A drink?"

Sheepishly, the boy shook his head and answered. "Well, French fry day is so rare…" He swiped a fry from the massive pile on his tray and devoured it in one bite.

Leena chuckled at the enormous mound of fries he had acquired. "I can't believe they let you have all those."

"I tried for more but they wouldn't let me. 'Three servings is more than enough, young man!' Sheesh. I could polish off a whole tub of them." He demonstrated by stuffing five fries into his mouth. Talking around them, he murmured, "You're welcome to have some if you're hungry."

"Both with apostrophes," Reiko said in an officious tone.

Nathan snagged a fry but commented on Reiko's statement before popping it into his mouth. "I guess you had Ms. Ettiqua this morning too?"

Leena cleared her throat and mimicked their English teacher with a warbly voice the woman didn't actually possess. "Remember, children, to use the *one simple rule* when you're spelling 'its' and 'your'."

Sergei snorted around a fry, laughing at Leena's interpretation of the teacher.

Reiko huffed, not liking it when anyone made fun of their elders. "It's a good rule." And because she knew Sergei well, she decided to remind him of it. "After you write the word 'your' go back and read the sentence as 'you are' and if it *does* make sense that way, then you *do* need the apostrophe."

Nathan laughed despite himself. "You even emphasized that rule the same way she did!" He then added the follow-up explanation, adding a British accent for fun. "It also works with 'its'. Say the sentence as 'it is' and if it *does* make sense then you *do* keep the apostrophe."

Sergei shook his head. "Nah, Leena does the voice better."

Reiko frowned. "Come on, communication is important."

"You're right," Nathan agreed, deciding not to upset her further. "But as helpful as that lesson was today, I'm really curious to see what happens in science."

Everyone started to smile and they leaned in conspiratorially so no one else would overhear them. "What do you think he has planned for today?" Leena wondered.

"No clue," Sergei replied, "but it sounded important from what he said Friday."

"Indeed," Reiko said, "he always has something wild ready."

"Remember what he said?" Nathan reminded them. "This whole week would be shocking. He challenged us not to miss any part of it."

As they entered Dr. Lupino's class at the end of the day, they were not disappointed. The wacky science teacher had covered every surface of the room with lights, lights, and more lights. Red and blue lights chased themselves across the side cabinets. Purple orbs encircled the chalkboard. Star-shaped bulbs framed the windows, and little LEDs even lined the tables and chairs. In some ways, it was blinding, but it didn't take long for their eyes to adjust.

But Dr. Lupino was nowhere to be seen. The twenty-eight students in his class settled into their seats, wondering how long it must have taken him to set up all those glittering lights. There was a 'Do Now' on the board, but no one even noticed it, least of all Nathan, whose seat was underneath a spinning disco ball.

Whispers echoed through the room and the sound waves all added together to make a great cacophony. It became so loud in the room Nathan thought someone would soon come in to yell at them for all the noise. Yet when the noise was reaching a crescendo, the lights all winked out for a split second and then they came back on.

And there, in the center of the room was Dr. Lupino.

The kids all gasped with shock. How had he gotten there so quickly? Where had he been hiding beforehand? And, why was his shirt all lit up with a pulsating rainbow of lights?

"Behold!" said the eccentric teacher. "I promised that you would be shocked, and so you shall. If you are not shocked already, I assure you… soon, you will be. Binders open!"

The magic words brought a universal groan, because it meant they would be taking notes. Dr. Lupino ignored the sound and he swept around the room, waiting until everyone had opened their binders to a blank page. Some students wrote their names or the date on the page, but Nathan waited. He didn't write notes until he was told to do so.

"Now!" Dr. Lupino shouted when there was a quiet lull from the shuffling of papers. "Something will happen today that you will not expect. For you will enter a new world. A different world.

And in this world you may or may not recognize anyone around you.

"But watch yourselves," he warned. "If you are not careful, you could be hurt. You must be cautious. Are you ready?"

Reiko raised her hand and waited until she was called on before asking, "Dr. Lupino, will this be on the test next week?"

He grinned. "Reiko, this *is* the test!" He raised his hands up and outward, like he was trying to catch an elephant, and the lights flashed off and on three times. Some of the kids looked frightened, but Nathan simply tensed, leaning forward in anticipation, wondering what was going to transpire next.

"When I say so, bend slightly forward. Then lean more. Keep bending forward until your forehead rests upon that open page in front of you. At that point, you will close your eyes. Count down from thirty and then open your eyes. You will be in a new world. Ready? Go!"

Nathan glanced around the room quickly and saw immediate compliance from most of the students. He shrugged and decided to play along. He leaned forward slowly, sneaking a peek toward Leena, whose head was already pressed against her page. He grinned, lowered his head, and closed his eyes.

Thirty, twenty-nine, twenty-eight…

CHAPTER 2

A NOT-SO-ORDINARY AFTERNOON

…Three, two, one.

Nathan opened his eyes and blinked several times, not believing what he saw. Gone were the lights and the desks. Gone were his classmates and teacher. He sat on a sandy boulder in a desert under a scorching hot sun. He shielded his eyes, trying to get his bearings. Nathan was utterly confused.

"Get down!" hissed a voice nearby and someone grabbed his arm and yanked him to the ground. He fell with a thud but he was told to hush when he moaned in pain. "Danger, there." A shadowy hand pointed toward the sunlight, and Nathan hesitated to look.

He didn't recognize the voice. "Who are—?"

"Hush!" the voice said. "No talk talk. Wait now."

Nathan didn't argue, wondering where on Earth he was. As he sat there in the sweltering heat, the ground underneath him shuddered for a moment then stopped. Then it shook again and stopped. The tremors repeated rhythmically and grew stronger with each passing moment. "What is it?"

A hand reached around and covered his mouth so he couldn't speak. He wriggled and tried to free himself, but his captor was stronger and held him firm. The tremors grew more intense and soon the light in the sky faded drastically.

Nathan saw a hulking form rise in the distance and stretch its neck into the sky. It blotted out the sun. From his vantage point, it was a gray, misshapen mass, but its size reminded him of one thing only: dinosaurs.

He glanced around at the rocky ground, the boulders, the sparse landscape, and the giant creature in the not-quite distance. It simply had to be a dinosaur! It was too enormous to be anything else. The creature continued its plodding gait and the tremors began again as it moved. Truly, it was a massive beast.

It felt as if hours passed by before the hand clenched around him released him. "Keep quiet. Follow." Nathan heard a slight scrambling sound and he reflexively ran after the person who had kept him from being seen by the dinosaur.

They ran for a number of yards. Maybe it was a mile. It reminded Nathan of gym class. He hated running the mile, but he always tried to push through the pain. It hadn't mattered to him that he wasn't very athletic, but it was hard to keep up with his protector right now.

Not long later, a few small huts rose into view and Nathan's host slowed to a walk and he was able to see who it was for the first time. In some ways, it reminded him of Reiko, but the person was very different. "Who are you?" he asked.

"I am called Treesa," she said. "This is home. Elk Tree Ville. You are visitor and unwary. You must beware."

"I don't understand. Where am I? How did I get here?" Nathan wasn't one to panic, but this was overwhelming. How could he have possibly ended up in a desert with a village and roaming dinosaurs? He wiped his brow nervously.

"You will know. Soon," Treesa promised. She smiled encouragingly and escorted him into the tiny village.

There weren't many people there at all, maybe fifty in total, ranging from babies to old folk. They wore dusty clothing that barely protected their skin from the sun. It made Nathan wish he had a bottle of sunscreen with him.

The huts around the village were made of the same beige cloth and it blended in perfectly with the tan color of the sand. At first he thought it was a rather boring combination and that it lacked any form of creativity. Then the ground shook with more tremors and he suddenly realized that the color scheme they had chosen made sense. It kept them hidden from the dinosaurs. Perfect camouflage.

Treesa walked Nathan through the village to the other side, which overlooked a wide chasm. "Protected, this side, from Walkers."

"Walkers?" he asked, then understood. "Oh, you mean those dinosaurs. You really live here with those things walking around?"

"One must live, yes?" she asked. "You know these creatures?" she wondered, pointing toward a great shadow that meandered along the horizon.

"Dinosaurs." He nodded. "Giant creatures that roamed the land in prehistoric ti—" He stopped himself. "Am I somehow stuck in prehistoric times? How is this possible?"

Treesa chortled at the expression on his face. "You make Treesa laugh. Those not called dinosaurs."

More tremors shook the ground as a small pack of Walkers wandered closer. Nathan noticed that there were two main varieties, but both were about two thousand times bigger than he was. One type of Walker was very dull in color and reminded Nathan of a gum wrapper after the gum had been chewed and spat away. That one seemed unimportant, even with its massive size.

The other type of Walker, however, was absolutely stunning. Its skin glimmered with shimmering light from every scale that covered its immense body. The colors swept up and around in a mesmerizing pattern and Nathan could focus on nothing else. He stared at the endless swirling effect and only the tugging hand and mocking laugh of Treesa pulled him back to the world around him.

"Those Walkers," Treesa explained. "Gray, more quiet, subdued. Those Neutrodons. Big, but no eat us. Squash us like ants sometimes, but no mean to."

Nathan was reluctant to ask, because he feared that if he looked again, the shimmering behemoth would entrap his attention again. Yet he had to know their name. "Those? The colorful ones?"

"Ah, yes, many go to doom when they meet Protosaurus. Lose self when they see bright color. Cannot resist pull of attraction." She grabbed Nathan's face and gazed into his eyes. "Keep distance or you not survive."

"I—I—" he stammered. Then he sighed. "I will stay away from them."

"Good. Is for Walkers to walk together. Not for us to be too close. We be Elk Tree. We survive by stay away. If too close, we eaten or squashed. We die." Treesa took his hand and led him from the area, bringing him to a campfire that was bravely set up in the middle of the village.

He had no idea what she served him for food, but it certainly wasn't French fries. It tasted a bit like rock sprinkled with sand and roasted over fire so the sand could become a little like glass. He didn't know how he got it down with only a few sips of water. Clearly, they didn't have any water fountains here and the water ration she had given him was barely enough.

A horn sounded in the air and Treesa tensed sharply, after which she placed her ear to the ground. "Tremors," she murmured. "They fade."

Nathan noticed all around the village that people were scrambling wildly. The tan tents were quickly collapsed and rolled into neat packets, and it reminded him of an umbrella that fell in on itself with the push of a button. Babies cried but their parents calmed them quickly. Although everyone moved rapidly, they did not move in panic. The haste was very different than panic and Nathan stood up, trying to understand what was happening.

Treesa grabbed his hand. "Come. Walkers move. We follow."

"What?" he asked incredulously. "I thought if we got too close, they would kill us or eat us or stomp on us or something!"

"Is true. But they go where is food. Our food, low. Need food. We follow."

Nathan scratched his head. "So what you're saying is that we need to stay sort of close to those massive Neutrodons and Protosauruses but that we also have to stay far enough away from them? We need them to survive but they'll destroy us too? That's crazy!"

"Is true," Treesa repeated. "And be wary of Protosaurus colors. Do not go near." She hurried away and Nathan pushed hard to keep up.

The entire village relocated over the course of several hours. They were constantly on the move, following the herd of massive beasts as they sought other nourishment. The sun sank low in the sky and then vanished altogether, revealing a flickering night sky, the likes of which Nathan had never seen. He stopped and stared upward in total awe.

"Is problem?" Treesa asked.

"I've never seen so many stars! Look how bright they are!" He thought of the bedroom he had as a little kid, where the floor was covered with pure black tiles. He had once purposely spilled some baby powder on the floor because it allowed him to glide back and forth along the floor. And he had for a few minutes before he had fallen and landed hard. The sparks in his head, mingled with the white flecks of powder on the floor, looked almost as dazzling as this brilliantly lit nighttime sky.

Treesa chuckled at Nathan's wide eyes. "Is story in village. Each light in sky is us from other side, looking down. Each tiny flicker is moment of life with meaning and power. Above, one light is Treesa. One light is Nathan."

It was an interesting idea. He stared upward at the glimmering lights. "I wonder which one is me?"

Weariness swept over him and he yawned. When he finished, he found himself standing upright in his science classroom, staring up intently at the spinning disco ball overhead.

Dr. Lupino grinned at him from the front of the room and Nathan flushed a deep crimson. He looked around the room but his other classmates were as dazed as he was. He glanced down at the blank page in his binder, shocked to find that it was no longer blank.

Inside there was a diagram of two larger objects, called protons and neutrons, and they were circled by other tiny objects called electrons. The words "electrons must never touch the protons and neutrons" were written in bold letters on the page.

Nathan thought back to the Walkers and their tremendous size, then he looked at the notes he had unwittingly scrawled across the page. He considered the warnings Treesa had given him about how he must never go too near the Protosaurus, even though he was drawn to it. That must have meant that he had played the part of an electron. Then at last he looked up at the teacher.

"Class dismissed," was all Dr. Lupino said, still grinning ear to ear.

ϟ ϟ ϟ

The bus ride home that day was overflowing with excitement. Nathan's classmates who shared his bus all sat together and discussed the images they had seen during class that day. Everyone had experienced a similar scenario, but each was unique. Though their stories were very similar, each was different because the person experiencing it had focused on different things.

"My guide," said Sergei, "was from Mother Russia, I am certain. She walked in the snow without heavy furs but wasn't cold."

"Snow? I saw a lot of water," Leena countered. "I was told to stay lined up on one of the rings of waves that came from the center and not to go closer than that."

Nathan scratched his head. "I wish Reiko was on this bus so she could help us put it all together. What about the notes?"

"So weird!" Leena nodded in agreement. "I don't remember writing anything, but there it was in my penmanship anyway."

"What does it all mean?" Sergei wondered.

Leena shrugged, opening her binder right then and there, it was so important. "Look! Protons and neutrons are in the nucleus and are almost two thousand times the size of an electron, which circles around the nucleus."

Nathan looked over her shoulder and read, "Protons have a positive charge and attract the negative electrons, but they must never meet or they would destroy each other. Neutrons are neutral and do not attract the electrons, even though they stay close to the protons."

Sergei turned from one to the other. "What does it all mean?" he asked again.

Nathan hesitated to answer at first, but then it gushed out of him. "The protons are those shiny Walkers I saw and all I wanted was to run over there, but Treesa said not to, that it would eat me. And I didn't much care about those gray Walkers either. It all fits! So strange."

"I'll tell you one thing," Leena said, lowering her voice and leaning in toward her friends. "Dr. L might be completely crazy, but I can't wait to see what happens tomorrow!"

"Me either," Sergei and Nathan echoed.

CHAPTER 3

TRACKING TUESDAY

Tuesday morning began like any other Tuesday, but this time Nathan was excited to head to school. He couldn't wait to see what would happen in science class. He and Leena endured English together. Today, partly because it was Tuesday, Ms. Ettiqua focused their lesson around the number two and other spellings of the word.

"To, too, two." She pointed to the words on the board. "With the 'w', it's a number, plain and simple. Think of two delicious sandwiches. There's a 'w' in the middle of 'two' and a 'w' in the middle of sandwiches. If that's not the one you're looking for, then look at the number of 'o's. 'Too' has too many 'o's to be serious. 'Too' means excess or extra or also…"

She carried on, but Nathan tuned out, daydreaming of the dinosaur landscape again, wondering what they might find today. Science class couldn't come fast enough. Nathan and his friends ate their lunches quickly, as if doing so would help them get to the brightly and oddly lit science room faster. Of course, they found the same thing that most people find: the more they wanted time to speed up, the slower it went.

Nathan and Reiko had social studies next. Mrs. Cofeni had them creating timelines. They needed to research a list of people—when they were born, when they became famous, and when they died—and then they had to compile a detailed timeline with the information. It required good organization to color-code everything so the data wasn't messy. They went to the library to find what they needed.

Nathan examined his list of people: Ampere, Einstein, Franklin, Faraday, Maxwell, Millikan, Thompson. Reiko's list was vastly different: Coulomb, Joule, Kelvin, Oersted, Ohm, Tesla, Volta. They were both a bit disappointed, because it meant they couldn't really work together when it came to researching these people, but at least they would be creating the final timeline together.

Finally, Nathan and Reiko met up with Leena and Sergei and took their seats in Dr. Lupino's room. The lights were all twinkling again, but there was something different. Leena pointed it out first. Most of the lights traveled in a large sweeping arc around the room, and it almost made the place feel like it was moving, sort of like the spinning Gravitron at an amusement park. Yet the door to the teacher's backroom had a different configuration of lights. They reminded Sergei of a chess board. None of those lights was flashing, but they were definitely in an alternating pattern of bright squares and dark squares.

"Binders open!" said Dr. Lupino from the back of the room. They hadn't seen him again, but with all the lights on his shirt, he probably blended in with the ones on the walls. This was the fastest that Nathan had ever seen a class open up to their notes. Binders snapped open and papers were flipped to an empty page. The kids looked around eagerly, wondering if they should lay their heads down and start counting.

"Reach down and grab onto your chair," Dr. Lupino instructed. "Keep a firm grip. Good. Now lean forward like yesterday. Touch your forehead to the page. Close your eyes, and count back from thirty."

Nathan wriggled with excitement as he complied with the instructions. He began his countdown, remembering to pace himself and not to miss any numbers along the way. The lower the count went, the louder a strange buzzing noise became, until it was almost unbearable.

He eagerly reached the number one and then opened his eyes, feeling a rush of wind against his head. He blinked a few times and saw to his surprise that he was behind the wheel of a racecar. And it was driving exceptionally fast! He glanced around and saw several other cars, all weaving carelessly across a strange metallic pavement. There was no one else in the car with him, and his reflexes took over.

He grabbed the steering wheel and held on for dear life.

The scenery whipped around him in a chaotic blur. He couldn't make out any real details, except that the sky was a very strange shade of green. It was the verdant green of leaves or grass, but it reminded him of the glow of a massive traffic light.

The road pitched to the left and he twisted the steering wheel so he could follow it. The racecar responded beautifully and turned exactly where he aimed it to go. The road swept back toward the right, then left again, and each time he conducted the car to follow. He peered outside his vehicle and saw that other drivers had also taken control of their respective cars. With the green-hued sky overhead, they all felt that the race was on! Go! Go! Go!

At first, Nathan was caught up in the exhilaration of the ride, but once he adjusted to the fast pace, he started remembering other things. For instance, the car was accelerating on its own and he had no way of stopping it if he wanted to. He risked a peek down to the floor but he didn't see any pedals. He tried not to panic, remembering only that Dr. Lupino had said to keep a firm grip.

With the green sky acting like a giant traffic signal, all the cars zoomed along the sweeping track. They kept riding really fast and it was impossible to see any details of the surrounding area, except the other cars that were going about the same speed and a sign that followed them that showed the name of the place: El Track Raceway.

Then all at once, the color of the sky shifted to an amber hue. As it changed, so too did the road they were traveling on. The metallic pavement changed to a gray, dirty roadway, which looked a lot like the graphite inside a pencil. Immediately the car slowed down. Then strange objects started to fall from the sky.

Nathan leaned heavily on the wheel to avoid colliding with a giraffe made of paperclips. Then a toaster oven dropped from above. Weird, random objects kept falling from the yellow sky and his car kept slowing down with all the obstacles he had to avoid. The new surface he was driving on also resisted the speed of the car and kept it from going its fastest.

He started to notice a strange scent in the air, too. It was akin to burning rubber. While he swerved around a falling tree house made of light bulbs, he peered outside the cockpit to the wheels. They were indeed smoking from all the friction! He was terribly worried that the tires might pop or something. He could feel heat coming from the tires, and they started to glow. Yet, there was nothing he could do.

Cries of dismay echoed across the field. It actually calmed Nathan to some degree, because it meant all the other drivers were having the same problem. He glanced up at the sky again. First green, then yellow. There was little surprise when it shifted colors again and turned a rich crimson.

As the sky went red, the roadway changed again. Instead of metallic pavement or dusty graphite, it suddenly turned into molten tar, which smelled a lot like melted rubber. It was terribly sticky and it caught the tires suddenly, stopping the car with a jerk. He had expected something like it, so he was insulated against getting hurt. His firm grip kept him safely within the race-car.

Looking around, all of the other racers faced the same situation. Their cars had halted and they could not move. He was tempted to climb out of the car and walk over to a nearby driver, but something told him not to budge. He noticed that someone else had the same idea, and he saw a silhouette jump down from the car and land on the sludgy ground. There, the figure got stuck. It was completely mired in the sticky goop and then it sank into the ground like quicksand.

Nathan was glad he had stopped himself from doing the same thing. He waited for the sky to change back to yellow or green so he could move along again. However, the weird racing world seemed to have other ideas. Instead, the whole place turned sideways. The drivers were dropped from their cars and they all landed in the tacky surface. Nathan's body was dragged downward, downward, downward, and there was nothing he could do to stop himself. Soon the thick goo rose up to his head and he took a deep breath and held it as he was pulled under the surface. Then, he was surrounded by total darkness.

He held his breath for as long as possible. His sight blurred and filled with sparks of lightning. And when he thought his lungs would explode, he let out a gush of air. Nathan panted, trying to catch his breath. His vision cleared slowly and when he could see again, he looked around the classroom. Sergei was lying on the floor, trembling in fear, but he was recovering. Apparently, he had been the driver to first go through the muck. The rest of the class sat upright, gathered their breaths, and then staggered out the door when Dr. Lupino dismissed them.

Reiko wandered over and collected her friends with a sweep of her eyes. Leena, Sergei, and Nathan all huddled together to discuss the events before the buses arrived.

"That was so fast!" Leena started.

"I cannot believe our incredible velocity!" Reiko agreed.

"I fell off the car. That was very scary," offered Sergei.

Nathan was scanning his notes, still unsure how he had even written them. "Did you all see the same stuff I did? Look at the notes; it all fits again."

Leena leaned over and nodded. "I guess a conductor is like a green light for electrons. It lets them flow through easily."

Reiko considered for a moment. "Yes, and the yellow light is like slowing down at a traffic light. There was resistance that slowed us down."

After that, Sergei sighed. "And that insulator... It did stop me from moving. Maybe it's the same for electrons?"

Nathan shook his head, bewildered. "What I want to know is, how is he causing us to see all that stuff?"

No one knew.

CHAPTER 4

DOWNHILL ON A WEDNESDAY

By Wednesday, all talk around the lunch table was about the crazy lessons going on in science class. No one understood how they were able to experience such adventures and then awaken to a set of notes, handwritten in their own penmanship. It was a cool effect not to remember the writing itself, but it unnerved them, too.

In gym class that day the students invented a game that reflected what they had learned during the racing adventure. They broke up into three types of teams, and the first two types took volleyballs. One type of team ran full-tilt like the conductors. The second team set was bogged down with padding, jackets, or by carrying extra equipment, acting as resistors. A few impromptu goalies stood in place, trying to stop the volleyballs from getting past them, like insulators. Anyone who was hit by the ball but didn't catch it had to become a different type of material. Thus goalies became runners or blockers became goalies, and so on. It helped keep the game entertaining.

After arriving in Dr. Lupino's class, the students noted that nearly all the little lights were white today. Gone was the wild spectrum encompassing the room. It felt brighter yet somehow it was calmer, too. On the board were written a few words in large, bold letters. "Slope. Bunny. Lift. Tree. Dove." Dr. Lupino stood nearby smiling but he refused to explain what the words meant.

The wacky teacher gave a slightly different set of instructions for today. Instead of holding on to anything, he asked the students to tuck their arms beside their torsos, with a fist on each side near their bellies. This was followed by the leaning forward, the closing of eyes, and the counting down from thirty.

Nathan thought he had gone blind when he opened his eyes. White light flooded into his vision and there it remained, even after he rubbed his eyes a few times. Blinking didn't help, either. Nathan reached his hands out to judge if anything was nearby, but he couldn't feel anything in the way. He turned to his right and laughed.

Off to the right side was a well-kept ski resort. He had opened his eyes facing the broad side of a snowy mountain! No wonder he didn't see anything but white. The ground underneath was a fluffy snow that didn't feel cold at all, especially considering that he wasn't dressed for skiing. In fact, he had never been skiing before. He hoped he could use a sled instead.

It didn't take long for Nathan to trudge through the snow to reach the ski resort, the Elk and Icy Tree. There were so many people inside. He searched for his friends, but he didn't think he would actually recognize any of them. After all, he hadn't really known anyone else in the other scenarios. The clerk at the desk shoved a set of skis in his hand and a series of large pictures on the wall demonstrated how to put them on and what to do with them when skiing down a hill.

A giant pulsating arrow nudged him to leave the lodge and to approach the nearest hill. He followed the sound of a dove as it sang "Coo! Coo!" while perched on a sign. He was happy to see that the sign called this the "Voltage Bunny Hill" because he knew those were the easiest ones to travel down. He wanted to watch a few people go down first and to see what to do, but someone bumped into him from behind and he was sent down the hill at unawares.

He knew a bunny hill wasn't very steep. And after the racing adventure, this was nothing in terms of speed. But he didn't have any protective gear on, so the speed at which he traveled terrified him. His arms pinwheeled around, trying to keep him upright. He wanted anything but to fall over or to crash into something or someone. Other skiers were traversing the hill with different amounts of skill, but he didn't dwell on that for long. He kept himself focused and alert for danger ahead.

Down the hill he bolted until he had a good amount of kinetic energy at the bottom. Then without notice, his body slammed into a giant sponge cake, making a minor impression. Moments later, another person sped down the hill and plummeted into the sponge cake right next to him. He was surprised to see that it was Leena.

His best friend fell to her knees gasping for breath. "Oh my, that was so fast!" She was just as startled to see Nathan as he was to see her.

"Fast?" he asked. "I couldn't imagine what an expert hill would be like if you thought *that* was fast." He held out his hand and helped her to her feet.

She looked at the impression he had made and the one of herself. She had plunged rather far into the cake. "I think we were on different hills."

Examining the depressions, Nathan had to agree. "I was on the bunny slope," he admitted abashedly. "Someone bumped into me and sent me down the hill, out of control."

Leena laughed. "Oops, that might have been me somehow. I accidentally shoved someone down the hill. But I followed right after, so how could we have been on two different hills? Oh, look out!"

She pointed up the hill as a skier approached them from the distance. Whoever it was, he appeared to be coming down the same hill that Leena had taken, but this skier was slaloming back and forth, as if avoiding obstacles. The impact with the sponge cake left an impression that was almost as deep as Leena's.

Sergei stumbled from the mushy padding. "No more trees!" he begged. "I swerved so I wouldn't hit them, and it slowed me down, but that was frantic."

25

Leena pointed as another swerving skier came down the hill, though this one went at a much slower pace. Reiko made the smallest impression of the four of them, for she had gone down the bunny hill and also had obstacles to avoid.

Nathan asked, "What do you think the importance of all thi—" But he was interrupted when the bench of a ski lift clipped him from behind and hoisted him back up toward the top of the hill. This time he could see several groups of skiers getting ready to go down the slope. One group had a large number of skiers. A second group had about ten. A third group had four, and because he had been separated from his friends, he suspected he would ski without a group. The dove was there again, still calling out, "Coo! Coo!" It was perched on a wood sign pegged into the ground that said, "This is the Current Bunny Hill." A horn blared in the distance and all four sets of skiers leaped forward, skiing fast down the hill.

Nathan felt the wind blowing against his face as he raced. The snow crunched slickly beneath his skis. If this was skiing, then he wanted to become a skier! He wanted to hit the slopes and ride the wind. There were no obstacles on this course and it was a bit steeper than the first bunny hill, but he truly enjoyed the ride. A fast peek to the side showed him that the other skiers were keeping pace with him. He felt like he was riding the current of a big water wave. They would all reach the bottom at the same time.

He could see the sponge cake approaching and he wondered what each group would do to it. He looked over at the other groups of skiers and decided that the more people in the group, the more they would squish the cake at the end.

With a squelch, all the skiers smacked into massive walls of sponge cake. Once they extricated themselves, Nathan went to examine the damage. It was as he thought; the group with the largest number of people made the biggest dent, even though they hit with the same speed. He wanted to discuss his theories with his friends but he wasn't sure he would see them again.

Still, he realized that the number of skiers affected the energy that was coming downhill, and from his first run, he understood that the size of the slope changed the amount of energy that each skier carried.

As he pondered this, the ski lift clubbed him from behind again and swept him back up to the top of the hill. The dove beckoned his attention, "Coo! Coo! Coulomb!" A new wooden sign boasted, "Resist Or Fall." He didn't like the sound of that. As before, some unseen force pushed him from behind and he went speeding down the slope.

He tried to forget that he had never skied before. He also tried to ignore the fact that there was a massive tree right in his path and he had no idea how to avoid it. The only thing that occurred to him was to tilt to the side. His body responded and the skis swerved slightly to the left and he went around the tree, but doing so slowed him down. He straightened out and continued down the slope, feeling the cold air whipping against his skin. A large boulder—then another—appeared ahead of him. His success against the tree bolstered his courage and he bent his knees, tucking his arms against his side and leaning right and then quickly left to veer around the stones.

Each time he turned, however, his skis felt a little warmer. When he completed several quick turns all at once, he thought he saw sparks flying off them too. He had no idea what to make of it, but the more he turned, the more light and heat his skis gave off. He wondered how much they could take before they broke.

This hill seemed to go on forever, and the number of obstacles increased greatly. The hardest ones to avoid were the giant polar bears because they blended in so well with the snow. Nathan did his best, sweeping himself left and right, hoping for that spongy wall to be waiting for him at the end, and soon. He was swerving so much now, his skis gave off a constant heat and they were glowing orange, sort of like the coils in a toaster or that thin filament inside a light bulb that made it glow.

The obstacles grew scarce and eventually disappeared altogether. Nathan was skiing down the empty slope by himself. He tensed, ready for impact with the wall at the end, but apparently this adventure wasn't over yet. He glanced around and saw hundreds of other skiers appear on the slopes with him and all around him. They were spread out fairly well so it was easy to keep going with the wide open hill, but then the walls closed in!

To Nathan's dismay, the edges of the mountain pressed together the farther they went downhill. The other skiers came closer and closer together, since they didn't have a lot of room to maneuver. Pretty soon, they were so densely packed that Nathan's elbow rubbed against another skier's arm. There was no way they were going to all make it through the canyon if they stayed together going the same speed.

Without ski poles or experience, Nathan wasn't sure how to slow down without throwing himself on the ground, which didn't seem like a good idea at all. People pressed in on all sides and it was a bit scary, as the walls ahead were so close it looked like only one person could pass through at a time. He leaned back and tried to let other skiers go. This move caused some of the others to bump into him and he felt the friction of impact with each and every skier who passed. They also bumped and jostled each other, unable to easily get through the narrow opening.

Like a mob of kids trying to leave the cafeteria, the skiers all slowed down while still trying to get through the narrow opening. Nathan wondered idly if that was what happened when he plugged in his TV set at home. Did all the electricity have a tough time getting through the wires?

It took some doing, but eventually everyone else made it through the opening. Next it was Nathan's turn. As he passed, he saw that all the other skiers had vanished. He didn't have time to wonder why, though.

At long last, he smacked into the sponge cake, after which he crumpled to the ground to catch his breath. For the most part, he didn't mind the skiing, but he hoped in real life that he would have more control over what was happening.

He waited at the bottom of the hill, curious about trying the slopes out again, but without the ski lift to get him up to the top, he was stuck down below. He tried walking up the hill, but some strange force that was different than gravity kept him from ascending. No, he needed the ski lift in order to keep skiing. Without something to put him at the top of the hill, his fun was over.

He looked around but didn't see the ski lift or his friends. He pouted, feeling suddenly bored. He faced the hill and peered toward its peak and hoped for the lift to appear and catch him and power his way up to the top once again so he could continue the skiing circuit.

At long last something made contact from behind and he sat back onto the ski lift, but it wasn't actually the lift. Instead, it was a chair, and he toppled over, staring up at the glowing ceiling of the classroom. He wasn't the only student in the class who had fallen over, so that made it a bit less embarrassing. He picked himself up, dusted himself off, and waited for Dr. Lupino to dismiss them.

"Some of you are showing signs of resistance," he said with a chuckle. "No problem, though. The week is halfway over. Very well then; it is currently time for you to head home."

It was such an abrupt ending that everyone simply walked out, dazed. Nathan didn't remember even getting to the bus or talking with Leena on the ride home. They just sat in silence, thinking about the skiing trip and trying to piece it all together.

CHAPTER 5

TRIALS OF THURSDAY

Nathan awoke a few minutes before his alarm clock and he started his morning, all prepared for the day ahead. His dad was in the kitchen making some pancakes, which was unusual for during the week.

"Here you go, Nate," his father said, placing a dish in front of him. "Fully loaded with chocolate chips, just the way you like them."

"Thanks, Dad! It smells delicious." He first reached for a container of orange juice and poured it into a glass, noting how easily it poured. He didn't know why it caught his attention until he poured some maple syrup onto the pancakes. The thick liquid slowly oozed out all over the pancakes, and he muttered aloud, "Look how slowly it's going. It has a lot of resistance."

"Sorry, son?" his dad said from the stove as he flipped a pancake into the air.

"Nothing, Dad. Just thinking out loud about something from school." Nathan swirled the orange juice glass around, wondering if he would always see things a little more scientifically or if it was just a phase. He took a few sips of juice and then polished off his pancakes before grabbing his things, saying goodbye to his parents, and heading for the bus.

Leena was awaiting him in her seat, beckoning toward him eagerly. "Come here, come here! Listen to this! I was telling my parents all about the skiing thing from yesterday and they offered to take us skiing this summer! Can you believe it?"

"Us? And, summer? Isn't it a little warm for skiing in the summer?" he laughed.

She punched him jovially in the arm. "Don't be silly. There are still places that are cold when it's summer time here. But yes, they said they spoke to your parents last night and they gave permission for you to go. Isn't that unreal?" She was practically bouncing off the seat, and not because the bus was going over potholes.

Nathan's eyes went wide when he realized that she was serious. "That's awesome! I didn't think I'd ever get to go skiing. My folks aren't real fans of it, which is why we've never gone. Wow, thanks Leena!"

She beamed. "I just hope I don't spend that whole trip thinking about voltage, current, and resistance with every trip down the hill."

"You too? I was thinking about resistance when I was eating breakfast. So weird. Oh, and you forgot about the ski lift."

Leena nodded. "Some kind of power source, right? It brought us back to the top and let us continue skiing. Hey, Nathan… Do you think that's what a battery is? I mean, everything this week has been connected somehow, right? And if you think about it, without a battery, something won't work anymore. Maybe the battery gives energy so those electrons can keep skiing—er, well, going through the circuit anyway."

Nathan agreed. "I think that makes sense. I sort of wish Dr. Lupino would explain this instead of trying to make us figure it all out on our own."

"Yeah," Leena chuckled. "Like what was the deal with that dove?"

They chatted the rest of the way to school and survived their first few classes of the day. Leena had math class right after English and when she met up with Nathan for lunch, she warned him, "It's algebra today with the Python."

Nathan grimaced. He hated algebra. At least the Python made it interesting. Actually, the math teacher was Mr. Gorrean and he made a joke when they discussed the Pythagorean Theorem that his first name should have been Python. It was an affectionate name the kids used when they talked about him—outside of class. No one used the moniker in front of him for fear of getting in trouble.

They stood on the lunch line, choosing between today's selection of vittles. The most welcoming of the ladies in the cafeteria pointed to a colorful side salad and said, "Home Slaw. Would you like some?"

"Home slaw?" Leena echoed. "It looks like a fancy kind of Cole slaw."

The woman smiled widely. "Why, it is! It has a few extra ingredients that are bound to give you a burst of energy. Very nutritious stuff, it is. It's my personal recipe," she said proudly. Both Nathan and Leena took hearty servings, partly because it looked interesting, but mostly to make the lady feel better.

Sergei and Reiko were having an intense discussion about the things they had been experiencing in science class that week. Leena plopped down beside Reiko and Nathan took a spot beside Sergei. Sergei immediately turned to him and asked, "What do you make of the dove from yesterday?"

With a shrug, Nathan answered, "I have no idea. All that 'coo coo' while perched on the sign. It didn't make any sense to me."

"Me either," Leena chimed in. "We tried to figure it out this morning, but all we came up with is that Dr. L is cuckoo!" They all laughed.

Leena was right about math class. Nathan and Sergei took their seats as Mr. Gorrean put an equation on the board. It wasn't a very complex equation, but it did involve some algebra to solve.

"Here is our equation for today: $V = I \cdot R$. If you have two of the variables, you can solve for the third one. The key is to get the missing variable on its own so that the other values will give you your answer." He popped open a marker and drew on the white board. "If 'I' is 4 and 'R' is 3, then you plug in and you can see that 'V' equals 4 times 3. And because the V is already by itself, you can just do the operation, which here is multiplication, and you get 12 as the answer."

To his credit, Mr. Gorrean knew this wasn't particularly interesting to everyone, but it was important to cover the basics. Nathan wondered how they could possibly make this more fun, like all the stuff going on in science class.

"Now, instead, if you have V as 10 and R is 20, then we have to solve for 'I'." Mr. Gorrean wrote it on the board: $10 = I \cdot 20$ He turned around and asked, "Sergei, how do we get 'I' alone?"

He couldn't resist. "If we all leave the room, you will be alone." The class laughed, but the Python merely raised an eyebrow and waited. "Sorry," Sergei apologized. "Divide by 10?"

The Python shook his head. "See here? It is 10 equals 'I' times 20. You need to get 'I' by itself, so you have to move the 20." He showed the work on the board. "It's like if you wanted to be alone, you would need to get everyone away from you. Does that make sense?"

Sergei nodded.

Nathan always knew when to divide, but he sometimes mixed up the number he should divide by. Even when he had that part right, sometimes he got the next part wrong; the division itself.

Mr. Gorrean knew it too. He set up the work on the board and pointed to it: $I = {}^{10}/_{20}$ Then he turned to Nathan to solve for I. "There you go, Nathan, what is the answer?"

He hesitated. He wasn't ever sure which part was divided into the other. Math just wasn't his strong suit. "Um… 2?"

The Python shook his head. "No, sorry. Think of it this way, if you have 10 pieces of cake and you cut them into 20 pieces, what size is each piece?"

"Half the size," he answered readily.

"Yes. They're not twice or 2 times the size. They are half or 0.5 times the size. You cut them into smaller pieces. Whenever you have a smaller number on top of a fraction, you need to get a zero-and-decimal as part of your answer. It's a number between zero and one because it's smaller than a whole piece. A whole piece would be 1. No piece would be 0."

He then wrote a new fraction on the board: $^{50}/_5$ "Now how about this one?" He kept his Python eye trained on Nathan.

Nathan looked carefully and saw that the top number was bigger, so he didn't think the answer would be 0.1 and then he thought about it in terms of cake. If he had a cake that was 50 inches long, and he cut it 5 times, he would end up with… "10 pieces!"

Mr. Gorrean ignored the 'pieces' part of the answer and said, "Yes! You've got it. Ok, one last example. What if you end up with a fraction that looks like this?" He wrote on the board: $^{60}/_{0.5}$

Nathan cringed. Fractions were annoying enough, but to put decimals in there too? It was just mean! He tried to think of it in terms of cake. He knew that 0.5 was a half. So if he had 60 pieces of cake and then he cut them each in half, he would end up with—he paused to do the math in his head—"Oh, it's 120."

"Perfect!" cheered Mr. Gorrean, after which he gave Nathan some jelly beans. "See? If you divide by a decimal, you get a larger value as your answer. It's basic math. Don't let it throw you."

He then turned the lesson back to the equation he had started with. They spent the rest of the period doing problems with $V = I \cdot R$. At one point, Sergei leaned over and asked, "Hey, Nathan? This equation. Don't you think it feels kind of familiar somehow?"

Nathan looked at it and let his mind open up a little bit. "Now that you mention it, it does. Like the V is Voltage. The R is Resistance. But I for Current? That's doesn't make sense."

"Maybe the Python and Dr. L planned this out. We should ask about it in science."

They waited until everyone was seated in science class before raising their hands and waiting for Dr. Lupino to call on them. He hadn't been answering questions at all this week, really, so they were a bit surprised when he offered an answer.

"You're very astute to realize that Mr. Gorrean was teaching you a formula that connects to the work we have been doing in our class this week. And you're correct to assume that V is for Voltage, R is for Resistance, and I is for Current. It is an 'I' not a 'C' because in the old days, it used to be called current intensity so the 'I' is really for the intensity of the current. It was used so much back then it just stuck and was never changed to 'C'.

"Each of those variables has a unit that it's measured in. Voltage is easy: volts. Current is measured in amps." He then drew a horseshoe on the board, saying, "And Resistance begins at Ohm. This formula is called Ohm's Law."

Leena nearly burst out of her seat. "Home slaw! From the cafeteria!"

Dr. Lupino laughed in earnest. "Indeed. And we show ohms with the Greek letter omega: Ω. So, are there any other questions or should we continue with our lesson?"

Everyone looked around and withheld their questions, wanting very much to dive into a new adventure. And just when they were ready for the next set of directions, Reiko asked, "Dr. Lupino? What was the significance of the dove yesterday?"

The teacher smirked and shrugged and then gave the set of instructions for today's journey. "This one is a bit tricky today; don't get rubbed the wrong way. Ok, now, lean forward... you know the rest."

Nathan was so intent upon asking his question at the start of class that he hadn't noticed the lights today. As he leaned forward, Nathan glanced at them briefly. The white lights from yesterday were dimmed a bit and looked sort of gray. Patches of the lights were out completely and he wondered if they had died or if they were out on purpose. He didn't dwell on it for long as his head touched the page and he closed his eyes, counting backwards.

The sudden sensation of floating filled him completely. He opened his eyes to see that he was hovering way up over the Earth, bobbing up and down in the atmosphere. At first he was afraid he would fall, but seeing numerous other people floating around assuaged his fears.

He swept his arms through the air like he was swimming and he was able to propel himself forward slightly. As he glanced around he could see large objects seemingly locked in place. He was drawn toward some of them, but some inner instinct kept him from touching them. It reminded him of the warning Treesa had given him when he had seen the dinosaurs.

Other classmates experimented with their limbs, swinging about and swimming through the air. A few of them came toward him and he tried to get out of their way. He needn't have really bothered. Some strange force repelled him anyway and as they came closer, he was pushed aside effortlessly.

Off in the distance he saw Leena and Reiko trying to come together. He swept his arms through the air and guided himself closer to them. As he approached, they all heard the dove as it flew easily through the sky calling its beloved cry, "Coo! Coo!"

"Nathan!" Reiko called in greeting. "Hurry, come here, but be careful or you'll push us away. We're all acting like big magnets or something."

"Don't magnets pull things closer?" Nathan retorted.

"Not if you bring the same poles together, they don't," Reiko answered calmly. "Opposites attract, not likes. Likes repel."

Leena started waving wildly toward someone in the distance and moments later, Sergei made his way over to them. He looked highly amusing as he did a backstroke though the air to reach them. It took a few moments of jostling around before they could hover close together without repelling away.

"Well this is fun," Sergei commented. "But I don't get it. All those other times we were traveling so fast. The cars, the skis. What's this about? We're just hovering here, not really moving at all."

Nathan looked around. "I'm not really sure, but I think that must have something to do with it!" His eyes widened as he pointed to a giant five-hundred-limbed octopus that was hurtling through the air toward them. Hundreds of furry tentacles writhed around, flopping aimlessly. The group of friends tried to hurry away before the octopus came too close, but there was nothing they could do.

The massive tentacles slapped forward and brushed against the entire mass of classmates and the odd structures that were just floating in the air. Nathan was surprised that he didn't actually feel any contact with the octopus, but something strange happened. As the wooly tentacles whipped back and forth, everyone was drawn forward.

From the furthest reaches of the airy playground, all the classmates were pulled toward the octopus like a vortex. It drew them closer and closer together, even though they all naturally tried to repel away from each other. The friction of the tentacles overcame the repulsive forces and all the classmates charged together. Then, at last, the octopus and its five hundred fuzzy limbs floated away.

There were many more people around than just the one science class. Nathan wondered if it wasn't the entire world gathered together. They just hovered there statically, unmoving. But they were also trying to repel away from each other and it felt like a piano was resting on his chest, the pressure was so great.

The large obstacles in the air didn't move at all, only these little tiny particles that they were. Then all at once, Nathan looked around and saw the Earth far, far below. He didn't know why he was suddenly so scared. Maybe it was the pressure squeezing him. Maybe it was because he could no longer move around. Maybe it was having so many others so close by. Whatever it was, the distance to the Earth below was unbearable. Leena was fighting back tears, he noticed, and Sergei's jaw was tensed in anticipation. Even calm Reiko looked frightened.

There was total silence as everyone held their breaths, afraid that the slightest disturbance would make them all fall to the ground below.

The dove swooped by, calling out, "Coo! Coo! Coulomb!" and its cry startled several of the people gathered together. It was enough!

The pressure crushing Nathan turned into a powerful, painful pull. It wrenched his heart forward and he plummeted toward the Earth. He accelerated so quickly he had no way of stopping it from happening. He screamed in terror as the ground rushed up to meet him. He wasn't alone. The entire population shouted in fear as they fell toward the Earth. Their collective scream must have been so loud to anyone on the Earth below.

They fell so quickly and so sharply that they left a trail of light behind them, which flashed briefly and then was gone. Nathan's body blasted toward the ground, finding a metallic tower that called to him. He tried desperately to aim for it. Maybe he could grab on and keep himself from hitting the ground. Leena and the others all reached out too. They hurtled through the sky, and the air molecules rushed in after them, beating a heavy drumming sound behind them. It rumbled ominously as they all yearned to catch the metal pole. Hands reached out. Hearts yearned for contact. Eyes focused sharply.

In a split second it was all over.

CHAPTER 6

FRIDAY'S FRONTRUNNER

Nathan had many nightmares of falling that night. He kept reliving the sensation of dropping to Earth from so far above. Once he even threw himself out of bed while he was sleeping and he woke up on the floor, sweating in fear. He reminded himself that it was just a dream and he climbed back into bed, hoping the nightmares would end.

"I guess that's what electrons feel like when they turn into lightning," Reiko said during lunch the next day. "They all build up, and then they suddenly blast to the ground."

Sergei shuddered. "I used to walk around the carpet in the living room, rubbing my socks on the ground and then I would shock my little sister after I built up a charge. I don't think I'll be doing that anymore. Not after experiencing that."

Leena snapped her fingers. "Static electricity. I get it. In the other scenarios, we were flowing along. We've always been electrons in these visions. We were flowing freely before, because that's what electricity is. It's electrons flowing in one direction! Don't you see? The race track? The ski slope? And then here, we all built up a charge and then once it was strong enough, it couldn't be held in place any more so it blasted us to the ground."

Nathan breathed deeply. "So if the electrons all gather first in one place, it's static. And then it jumps? I guess that does make sense."

"Oh!" Sergei gasped. "And that monster that rubbed against the cloud, that's like me rubbing my socks on the carpet. But what were those structures floating in the air with us?"

Reiko had the answer. "If we're the electrons, then those have to be the protons and neutrons, don't they? They're too big to move around, but we're tiny electrons so we can move freely from one place to another."

It was all starting to come together. Indeed, in that first moment with the dinosaur-like creatures, they were also electrons. Each scene thereafter kept them in that same role.

"So electrons go down the ski slope," Nathan said, trying to put it all together, "and the steeper the slope, the more energy they have at the end."

"Yeah," Leena agreed. "Different voltage. You get more electricity out of a 9-volt battery than a AA battery, and even more if you plug it into the wall."

"And without the power source, or ski lift, the circuit stops working," Sergei added. "So I guess the power source gives energy to the electrons so they can keep traveling?"

"That would also explain why batteries die," Reiko offered. "If they give their energy away, they have to run out some time."

They were so caught up in their chat, they didn't hear the bell ring and were almost late to their next classes. When they met up again in science, they all took notice of the lights. Once again, they were lit and chasing each other around the room, but they were different, too. There were strange patterns, but they didn't have time to ponder them.

Dr. Lupino had written a negative number on the board in scientific notation. -1.6×10^{-19} The number didn't have any significance, so they just looked at it, figuring it must be important somehow. The teacher didn't explain the value, but gave them today's warning, "Be wary of the path you choose. Make sure you always return home by the end of the day." Everyone set their heads down to begin.

Nathan opened his eyes and was perplexed. He was standing on a street corner looking into a store window. The things happening inside made no sense. He watched a girl make a cup of coffee, but she was doing it all wrong.

First, she poured milk. Then she took out a mug. She sipped from the mug, but it had to be empty because she hadn't poured the milk into it. Then she turned on the coffee maker, after which she added sugar to the mug. She put coffee grinds into the coffee maker then she stirred the contents of the mug.

Nathan was highly confused and turned away. It would have made a lot more sense if she had put the grinds in first, turned on the machine, taken out the mug, poured the coffee and milk, then added sugar, stirred it, and drank it. It was silly otherwise. He walked toward another store and saw someone making what should have been a peanut butter and jelly sandwich in much the same way; completely out of order. It was hard to put peanut butter on a sandwich if the bread was still in the cupboard!

He didn't like these events happening out of order. Things were meant to go in a certain series in order to make sense. It was like watching a TV show. If he missed a few episodes, he didn't know what happened to the characters and he felt lost. It all needed to go in order. One way. He glanced at a sign over the town: "Welcome to Electri Town." He grumbled, not feeling particularly happy being there.

Nathan grabbed his head to try to get the images out of his mind. He then looked around the rest of the odd little town. The street was lined with numerous inflatable punching bags, the kind kids bop and they bounce back up to be hit again. He walked over to one and punched it. It dutifully went down and came back up for another beating. He played with it for a few minutes before realizing that he needed to move along.

Yet as he stepped away, the punching bag leaned forward and bopped into him. He turned back and punched it down again. Then the next one started rocking, and he dodged out of the way and then knocked it down. Peering down the road, he understood. He needed to knock them all down. Gleefully, he sprinted down the street, punching left and right, knocking down all the punching bags and running around the block until he ended up at the start again. Each one slowed him down a little bit as he went, but all the bags remained down this time, so when he returned to the start, they were all pointing the same way and he was very tired.

He paused to catch his breath and saw that a small table was nearby with a sandwich. He scarfed it down and felt empowered to do the run again. The punching bags all rose back up, ready for him. He went off again. Five, six, seven punching bags all went down easily and then something happened that he did not expect. A giant chasm opened up ahead of him. It was way too large for him to jump over. He couldn't finish the run, so the punching bags on the other side were allowed to just stand there, untouched. Disappointed, Nathan trudged back to the start.

He refueled with another sandwich, hoping the chasm would close itself just as quietly as it had opened. But then he heard a loud grating sound. Something large had moved somewhere. He would have to keep an eye out for it.

With a grin, he ran off again, belting the punching bags as he went. Each one he hit slowed him down a little bit, like before, but he didn't care. This was a good way to relieve some stress. At the end of the row he turned and that's when he understood what had changed. The path ahead of him split in two and he could see punching bags lining both roads. He didn't think he would have the energy to go down one road and then come back to go the other way, yet he felt compelled to knock all the bags down. Instead, he wished he could go both ways simultaneously, and so he did.

Somehow, Nathan's body split into two identical bodies. He glanced at himself and it was like looking at a twin. He laughed at himself and then both Nathans ran ahead, each one taking a different path. They each took care of their set of punching bags and when the paths came together again, there was no need to backtrack at all. The two Nathans raced for the start line, joining back into one single Nathan. He was twice as tired as before and it took more sandwiches to rebuild his strength.

On his third run, Nathan expected there to be even more paths. He wasn't disappointed. When he reached the branch point, Nathan split into four likenesses of himself, which he affectionately called N, A, T, and E. "I'll see you guys at the finish line! Don't keep me waiting!" said N.

"Not a chance!" replied A.

"See you there!" agreed T.

"Let's do this!" cheered E.

And off they went. N sped around the corner, knocking down all the punching bags in his way, rounding the bend and heading for the home stretch. A took the second parallel path and had no troubles reaching the end of the pathway.

T, however, found that his pathway was cut open by another chasm, like the one from before. There was no way he could make it over the hole, so he went and followed E's path. All four met up again at the starting line and they fused together back into Nathan.

He was ravenously hungry and he devoured several sandwiches that time. He was lucky that there had been another path for T to take, or else he might have lost a part of himself. It was unfortunate, though, that they couldn't knock down all the punching bags, but at least they took care of most of them.

While he ate, he thought about what was happening. When there was only one path for him to follow, he had to follow it in order, going just the one way. Each obstacle slowed him down so that he was tired by the end of the run. When the chasm had first appeared, he wasn't able to finish the run at all. He thought back to his conversation with his friends at lunch. If he was an electron, and this was a circuit of some kind, then interrupting the circuit would stop it from working. It was like making the coffee or watching a TV series. Something in series had to go in one direction only for it to work.

However, in the later runs, he had more than one way to go. It took more copies of him to go down all the paths, but they were able to split up and do it anyway. It was more of a drain because more electrons were going through, so more current. And if one pathway was interrupted, then those electrons could go down another path and everything else would still be okay. Just the one path wouldn't work correctly.

He thought of holiday lights, where if one light went out, the whole string stopped working. He guessed that was like a series circuit. On the other hand, he could flip a switch at home and turn on just one lamp in his house without having to turn every-thing else on too, so that must be connected in parallel.

As he realized this, the dove swooped in. "Coo! Coo!"

"What are you?" Nathan asked.

"Coo! Coo!" replied the dove. "I am you! Coo! Coulomb! You have a charge! Coo! A tiny number! A tiny particle!"

"I thought an electron had a charge of negative one," he argued.

"Coo! Coo! That is true! But only if you're big like me and you. Coo! Coo! It's easy for you to use a number like negative one. But coo! Coo! Coulomb! You have a real charge, in Coulombs! Coo! Coo!"

"A real charge, in—"

The dove suddenly flew off and as Nathan watched it fly into the sun, he squinted against the light and when his eyes cleared, he was back in the classroom, staring at the number Dr. Lupino had written on the board: -1.6×10^{-19}

"—Coulombs," he finished saying.

CHAPTER 7

DR. LUPINO'S DOMAIN

Everyone looked around the room, having woken up suddenly. The light patterns along the walls, desks, and ceiling had changed and were fluctuating randomly. Dr. Lupino was at the back of the room, holding his head tightly.

Nathan scrambled out of his seat and he rushed over to the teacher. "Dr. Lupino, what is it? Are you hurt?"

"I—I was attacked," he gasped. He truly looked distraught and in pain. "Everyone, I need your help. I know there is little time left in the period, but I need you to all go back in. There is a great battle, and only you can help turn the tide! Please, I beg of you."

Some of the kids in class looked scared at this declaration, but Leena stood up, and so did Sergei and Reiko. "You got it, Dr. Lupino. Right, everyone?" she asked, sweeping her gaze around the room. "Whatever it is, we can do this if we do it together!

Right?" A few kids agreed, so she shouted louder. "Right?!" The entire class chorused in agreement.

Everyone settled back to their seats and Dr. Lupino made his way to the front of the room. "There is so little time. Be safe, all of you. Go!" As one, the class flipped to a blank sheet in their binders, after which they lowered their foreheads, then counted back from thirty.

Loud whistles echoed through the air and Leena grabbed Nathan and pushed him aside as some sort of projectile flew toward them. He looked up into a cloudy sky, nervous with all the noise. His entire class was huddled nearby and he could tell that they were all here together this time. Sergei scouted the area and reported back that enemy forces were approaching quickly. Some of the kids screamed in panic, but Reiko calmed them down.

"What are we facing here?" Nathan asked, but no one knew. "Ok, then we have to figure that out, too."

Reiko turned to him. "Nathan, we can't all stay in one place. We have to move. We're easy targets if we just remain."

"You're right," he agreed. The decision was made. Each of the four friends took six of their classmates and led them off to another part of the area. As he led his troop, Nathan saw that they were in a sort of military compound. The location was spray-painted on one of the walls: Electri City. Sandbags bolstered the cement walls, which were lined with barbed wire. The thumping sound of an approaching army echoed over the stony barrier.

"Incoming!" Brian called from behind Nathan. Overhead, a blazing ball of energy flew over the wall and impacted the ground with a rattling force, shaking the kids to their feet. "Another!"

Nathan looked around quickly. There was a random smattering of supplies everywhere, as if this fortress had been quickly abandoned in the past and no one had even come to loot it. He scrambled around on his hands and knees and he found a rubbery tarp. "Everyone, here, help me with this."

Nathan's group made its way over. Each child took a part of the giant tarp in hand and Kevin, usually so quiet, asked, "Now what?"

"Lift it up! Those are electrical charges they're sending in. This is rubber. It's an insulator. Remember the race track? The sky went red and the road turned to rubber and we stopped moving. This will stop the electricity."

"Are you sure?" asked Brian.

"Lift!" Nathan implored. As one, the students all raised the tarp up high as the next electrical volley shot over the wall. The whirling ball of energy plummeted down toward them and struck the tarp with great force, shaking them, but then it dissipated completely. It had worked!

Nathan called out over his shoulder, "Leena! We need you here!"

Moments later Leena's team joined them and, while Nathan's group kept the tarp hoisted up, Leena's pack scrounged around for makeshift poles. From various plastics and tarp, they created a sort of rubbery tent, where they could plan their next move.

Leena sprinted off with her troop to help the others set up similar defenses, while Nathan huddled around with his classmates to discuss what to do next. "It isn't enough to stop the charges," Nathan said. "The army is still coming. We have to fight back somehow."

"Why would Dr. Lupino put us in this kind of danger?" Taylor queried. "Does he really think we can survive all this?" A blast of energy impacted the outer wall and parts of the stone cracked. It felt like the army would burst through at any moment.

"We can do this," Nathan declared. "We need to redirect the energy toward them."

"We have no way of doing that!" Jacinda said.

While Nathan considered, Sergei ran over to his shelter. "Nate, that last blast… it electrified the barbed wire over the wall. That wire's a conductor. It's getting really dangerous in here now. We can't even try to climb over the walls or we'll be fried."

A ball of lightning flew into the encampment and exploded one of the small shacks. The iron pipes that were inside rained down dangerously onto the field. Everyone dodged out of the way and luckily no one was hurt.

But it gave Nathan an idea. "Get those pipes over here, now!" Everyone within earshot heard the command and they hurried to comply. Soon, over a hundred iron pipes laid piled up at Nathan's feet. Each one was slightly tapered on one end and they were hollow. The tapered end would allow them to connect the pipes together, sort of like connecting two straws after pinching one end.

"What are we making?" Sergei asked as he helped put pieces together.

"A really big lightning rod, to draw the energy away," he answered. They combined several sets of rods into poles that were five times taller than the kids were and they tied them together into a thick bundle with some rope they saw lying on the ground.

They waited until an energy ball sailed overhead, then raced into the center of the camp and jabbed one end of the lightning rod into the ground. They worked quickly to drag some bricks over to support the pole so it would remain upright on its own.

They scattered away, and none too soon, for another energy ball flew toward the base and changed direction in midair, impacting the new lightning rod and cascading safely into the ground below.

"This protects us, but it doesn't stop the attacks," Sergei said. He ran to a hole in the wall and peered through before coming back. "There are hundreds of them in armor and everything! We can't fight them off like this. We're running out of time. They'll be here soon!"

Reiko and Leena joined them, now that the others were relatively safe. "We don't have anything we can use to make a catapult," Reiko announced. "Not that I want to actually hurt anyone today, even after all these volleys. We have to find another way."

Leena agreed. "Remember when we were in that cloud? As static electricity? We were all electrons, all charged with the same polarity. Can't we do something to repel their attacks and send the energy back to them?"

"It's too powerful an energy ball for that," Sergei denied.

"Not if we all work together," Nathan decided. "Leena's right. We were all grouped together in one place before we shot out like a blast of lightning. What if we tried it here?" They had no other ideas, so they each called to their group and summoned the entire class over to join them.

Timidly, the twenty-eight students stood in a huddle near the lightning rod, waiting for the next attack and hoping they would be protected. They held onto each other tightly, some defiant, some terrified, but all unified against this assault. They looked upward as the next glowing orb entered view. It veered toward the lightning rod and ran down the iron tubing, but Sergei noticed that something was different. The lightning ball jumped forward at one point and when he looked around at the group, he thought he knew why.

"Everyone, quick! Line up and face the same direction! Yes, look that way and stand in a grid pattern. Line up! Line up!" Everyone did as they were told without question and the next volley shot over the wall. "Hold steady!" Sergei shouted, sweat lining his brow. He hoped he was right.

The ball surged forth and turned toward the lightning rod, but then something strange happened. The ball hit some kind of invisible field and was repulsed to the ground, forward and down. The next blast of energy did the same thing.

"Everyone, turn around!" Sergei called out, needing to test his theory. This time when the ball struck the invisible shield, it was sent arcing upward and over them, and then into the ground ahead of them. It was like they had a massive circulating field around them that carried the energy when it touched.

"Excellent work, Sergei!" Nathan cheered. "I guess we are still acting like electrons here after all."

"I don't understand," said Leena. "What is happening?"

"We're like a magnet!" Sergei beamed. "And we're being protected by a magnetic field. But it only works if we're all lined up like this. That's what I saw before; the few of us who were facing the same way deflected the lightning ball as it ran down the pole. Now we're all working together and protecting ourselves. This is our domain now!"

The next energy blast hit the magnetic field and was deflected onto the barbed wire along the stone wall. They could hear the energy crackling all along the edge, and then a mild clanging sound echoed. Nathan looked back and saw some nails and bolts sticking to the iron lightning rod, but then a few seconds later they fell to the ground.

"Everyone, try to make the energy hit the fence again," he called.

As one, the class stepped forward to help deflect the next energy bolt in the desired direction. They were successful and the clinking sound happened again. Nathan turned and saw some metallic objects flying toward the iron pole and then moments later falling off.

"It's a temporary magnet," he decided. "When the electricity goes around it, the iron in the middle becomes a magnet, and when the charge is gone, it goes back to normal."

Suddenly, the sound of a battering ram impacting the walls rang through the air and everyone screamed in panic. "Hold still!" Nathan implored. "We need to keep up our defenses. We can do this, everyone. Keep it together!" The electric charges kept flying overhead while the other members of the army battered down the wall.

Reiko called Nathan's attention to a pile of copper cookware and pipes that she had seen in one of the remaining storage huts. "We could use that to channel the energy better than the barbed wire," she offered. "Maybe we can make this magnet in the middle really strong!"

Several kids ran off with Reiko to fetch the materials. This weakened the magnetic field but they were still able to turn the energy bolts aside. They didn't know what they were going to do about the army breaking through the wall, but they could only handle one task at a time right now.

Even better than the pots and pipes, the group found full sheets of copper paneling. Each was almost as tall as they were and they carefully created a wall encircling the iron rod in the middle of the camp. Meanwhile, the rest of the class redirected the energy blasts toward the army that was still breaking through the wall. Once the copper barricade was set around the iron, the next part of the plan came into play.

Nathan called for some volunteers. "Who is the most athletic here?" Brian, Taylor, and three others raised their hands. "Great. I need you to jog around that copper wall for as long as you can. Spread out and keep moving."

"I don't get it. Why?" Brian asked.

"We're still electrons here," Nathan explained. "You will be electricity. It will make that center piece of iron into a magnet… a magnet powered by electricity."

"An electromagnet," Leena smiled.

Nathan agreed. "As long as you keep running, it will be a magnet. And remember when we skied. If you can get more of you to run, that'll make a stronger current too. We'll take care of the rest."

"I don't see your bigger plan, but we'll do it," Brian acknowledged, and off they went, running around the copper wall. Moments later, small clinking sounds could be heard as the nails and bolts flew back toward the central iron column. It was working.

"Now what?" Sergei asked.

"I think the army has been using static electricity to generate those lightning blasts. If they can do it, so can we. But we also know how to deflect them. That means, we have to make sure they don't line up neatly, or else they will be able to block them too." He glanced over his shoulder as Taylor passed by, running at an easy gait. "I hope they can keep that pace for a while."

"They're the best runners on the track and lacrosse teams," Reiko reminded him. "They'll be fine. What's that magnet for anyway?"

Sergei gasped. "The army! They're wearing metal armor!"

Nathan only grinned in response. The hammering at the wall continued, as did the lobbing of energy blasts over the wall. The rest of the class still deflected the energy as best they could, sending it to empower the barbed wire that encircled the camp.

At last, the stone wall broke apart and members of the army spilled into the area. Nathan grabbed seven of his classmates and shouted, "Huddle! Quickly!" They pressed together as closely as they could and when Nathan felt a pressure build in his chest, he took a half step forward, disturbing his equilibrium. Instantly, he blasted across the field in a streak of lightning and knocked down three of the invaders. His body snapped back to the rest of his group and he looked at the damage he had done.

The three army men had fallen and immediately the magnet in the center of the camp caught their nickel-plated armor, lifting them from the ground and summoning them to the central iron rod. The nickel in their armor added to the magnetic power of the iron core and so it became even stronger.

The classmates took turns now setting up a magnetic field by lining up in a distinct pattern, and then huddling together to act like a static charge. At times they lined up as one strong, consecutive force, sort of like a series circuit, each giving strength to the person in front. Other times they had to split into smaller parallel groups and use more of their individual energy to fight back.

Little by little, they whittled the army away. The last three invaders wore dark blue suits of armor made of cobalt, but the material was also strongly attracted to the magnet and so it offered them no defense against the attack. Finally, the assault was over.

The five students running around the copper barrier were exhausted, but Brian kept encouraging them onward until further instructions came their way. The iron core in the center was covered with members of the invading army, all pressed together by the magnetic forces at work. It was like they were the nucleus of a huge atom, clustered together like protons and neutrons, being circled by a set of electrons.

At last, the runners could run no longer and they collapsed and rolled away to safety. Everyone feared that the army would fall to the ground and retaliate, but they soon learned something else. Iron, nickel, and cobalt were apparently very good at holding a magnetic charge, so even when the electricity running around the outside had stopped, the magnets held true. With the armored soldiers clustered together in the middle, Nathan thought they looked like a swirling, magnetic mass. He remembered from an old science lesson that those three same materials of cobalt, iron, and nickel were in Earth's core and generated the magnetic field that surrounded the planet. But these soldiers were stuck and un-moving.

The students had won!

The class cheered at their victory and they all ran together, creating one last supercharge of electricity. Then, like fireworks, they blasted into the sky and illuminated all the world below.

EPILOGUE

WHAT IF?

The class woke up and cheered as one. They had worked together and used everything they had learned through the week and they had held off the enemy forces. Even Dr. Lupino was impressed.

"Yes, yes! Well done, all of you!" he commended them. "I should never have put you in such danger, and I am sorry."

Nathan looked around the room and saw that all the lights were off. "Dr. Lupino, what happened? How'd all this occur?"

He looked around at the class, seemingly nervous to answer the question, which was very unlike Dr. Lupino. He usually answered everything willingly, unless he had a reason for holding back his answer for a time. "Well, all right. After what you have endured, I suppose I ought to explain some of it."

The class didn't even care that the period had ended ten minutes ago. They wanted to know the reasons behind all of this.

"Do you know at all how the brain works?" he asked. "Simply put, it's this: The brain uses a wide array of electrical charges to carry information from one end to another, carried across neurons. The body does the same all throughout. Electrical impulses. Electrons, bouncing from one place to another with purpose. It's one of the reasons we can be easily electrocuted."

He cleared his throat. "What if you could control all of those impulses? What if you could make those electrons all go in one direction at will?"

Leena called out, "Why, you could generate your own electricity."

"Precisely," the teacher agreed. "And you also know that one electron can affect another. They repel each other, and by using that, you can guide them where you would like them to be."

He took a sip of water and then continued. "So I wondered if I could direct my own electrons and thereby influence yours."

"You used mind control over us?" Sergei gasped in shock.

"No, not quite, no. But I did use a little psychology. Have you ever seen a magician perform a magic trick in front of you? You expect a little magic and so when he or she does the trick, you see the magic, not the trick. Unless, you're looking for the trick, that is." He tapped his brow. "Remember what I told you all last Friday?"

"Yes, that this week would be something not to miss," Reiko answered. "And that it would really shock us, by the end of the week."

"Indeed. And weren't you all expecting something completely amazing?" The class nodded. "And then all these lights?

Well, they were set up to illuminate in very specific patterns that reflected the way I wanted the energy in the room to flow."

Nathan hummed in thought. "So each of us felt the energy in the room and were sort of swept along with it, and you were guiding us the whole time."

"Yes," Dr. Lupino affirmed. "It wasn't hypnosis, but it was a lot like it. The difference was that as you felt the energies in the room, you were counting down with your heads on the page. Each day, I started reading a story to you at that point. And each day, you lost yourselves in that story. Your imaginations got the better of you, in the end."

"But our notes!" Reiko interjected.

"You wrote them while you were daydreaming," the teacher explained. "I merely gave you the basic structures of the story and your minds did the rest. Incredible device, the mind."

"So what happened in the end with the war?" Nathan asked.

"Ah, yes. You see, as the week went on, you each resisted the pull of the illusion more and more. It is why some of you started seeing each other in your stories. By today, you had all discovered some underlying principles at work and that brought you closer to reality. When I implemented today's story, I was anticipating some backlash and I received it. In a sense, I created it by expecting it, just as you created your worlds by expecting them."

"Too wild," Sergei said, shaking his head slowly side to side.

"Yes," Dr. Lupino agreed. "But I think it sends an important message to us all. Our thoughts and expectations empower our lives. The things we think can become reality. If we think bad things, then bad things will come to pass. If we think pleasant thoughts, then we create those as well."

Leena considered it for a moment. "Sort of like, if you do good things for other people, then good things will happen to you."

"Yes, and not necessarily when you expect them to, mind you. I have seen through this experience with all of you that the mind has untapped power and we should all be wary of it and conscious of it too.

"So I leave you with these things to consider:

"What if we could harness the electrical power of our minds: would we be able to create real magic, like in the mythical days of sorcery? Would we be able to cast spells and shoot lightning from our fingertips? What kind of spells would you want to learn?

"What if we harnessed our minds and used those powers to positively influence the world around us? How would you influence the world around you?

"What if we could change the world with a single series of thoughts? What would you do with such power?

"Now ask yourself: What if you already can do these things and just don't know it yet?"

⚡⚡⚡⚡⚡

AFTERWORD

There are many key aspects of electricity and magnetism discussed in this story through analogy. Before originally writing this tale in May 2011, I used these analogies to help students visualize the scientific concepts while discussing the terms in class. I have also noticed over the years that fewer students seem to read for fun, as added pressures and challenges take free time away from their lives. I decided to put the two together and thus *A Shocking Journey* came to fruition.

Here are some fun facts about why certain names and themes were chosen:

Reiko was a friend of my sister's way back when I was kid. Reiko was a foreign exchange student from Japan and she was a bright child with a shining personality.

Sergei's name was chosen because the theme of this story is electricity, and it's based on the word *surge*, as in power surge.

Dr. Lupino was named after me, to be honest. A wolf is known as *canis lupus* and so it's a play on that. Also, I have a PhD in science education, so it made sense that this teacher would too.

Ms. Ettiqua, the English teacher, received her name after the concept of etiquette, which is a set of societal rules for proper behavior. You know, like saying "please," and "thank you," or knowing which fork to use in a restaurant, and so on. I included the grammar lessons because Reiko was right: communication is very important in life and we should all do our best to be as clear as we can be. With social media and text messaging, sometimes people forget the basic rules for grammar and the meaning of a statement can change for the worse. Have you ever had a text message misunderstood?

The social studies teacher, Mrs. Cofeni, provides a reminder of the three strongest natural magnetic elements, which are also in Earth's core. The letters of her name come from the chemical symbols for cobalt (Co), iron (Fe), and nickel (Ni), as they're found on the periodic table of elements. All the scientist names that Nathan and Reiko had to research for their timeline are people whose work led to our understanding of electricity and magnetism today.

The math teacher's name was explained directly in the story. He was named after the Pythagorean Theorem. Most people know it as $c^2 = a^2 + b^2$. A lesser known fact is that the Pythagorean Theorem is actually a special case of the Law of Cosines, which is: $c^2 = a^2 + b^2 - 2ab \cos C$. The Law of Cosines works for any triangle. If you're working with a right triangle, there's a 90° angle in there. The cosine of 90° is zero, so that last term disappears and you're left with the Pythagorean Theorem, which only works for right triangles!

I also threw a bunch of electricity and magnetism puns in there. You probably caught a handful of them, but a few were subtle. Toward the end, when Sergei says, "This is our domain now!" he is actually referring to a magnetic domain, where the electrons spin in the same direction to create a magnetic field. Can you find some others?

Acknowledgements

I am deeply grateful to all the students I have worked with since beginning my teaching career. At every turn the curiosity, excitement, and humor of my students have inspired me to do all I can to best teach them.

I thank Rochelle Deans for taking the time and effort to professionally edit this story. It wasn't easy making some of the changes, but you were right. Doing so made the story much more enjoyable without losing any of the goals I had in mind. (www.rochelledeans.com)

I am grateful to Matt Caulkins for the awesome work he did on the cover. Trying to incorporate seven images into one was no easy task, but Matt made it happen like it was no trouble at all. (matt-caulkins.squarespace.com)

This story has essentially had more than 1,500 beta readers over the years. Each time, students have reported back on their favorite parts, and the parts they felt were confusing. I've taken each piece of feedback and worked with the story to make it the best it can be.

I also want to give a special shout-out to Taylor for taking the time to read and comment on the tale. You offered another perspective and I appreciate it.

To all my family and friends, you've always been the chapters of my life. You've built me up. You've warmed my soul. And I am eternally grateful to you all.

ABOUT THE AUTHOR

 Stephen J. Wolf earned his PhD in science education in 2006 and has worked as a middle-school science teacher since 2001. His passion for chemistry and physics was inspired by watching *Mr. Wizard's World* as a child and learning that many of life's biggest mysteries could be explained through science.

When he isn't helping his students discover logic and wonder in the classroom, Wolf enjoys spending time with his partner, Kevin, and watching *Doctor Who* with their two cats, Merlin and Monty. Wolf is the author of the *Red Jade* fantasy book series (tinyurl.com/redjadebookseries). He currently resides in New York, and you can visit him online at StephenJWolf.com.

www.ingramcontent.com/pod-product-compliance
Lightning Source LLC
Chambersburg PA
CBHW071345130626
46556CB00005B/2033